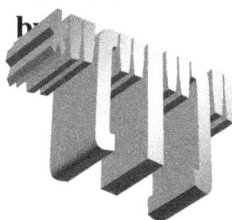

*"((bLONDe_bOX) Theory of reCONSTRUCTION)"

by
Stanley Alexander MARTIN/Nana baBa jaH-aYe

…++"{….

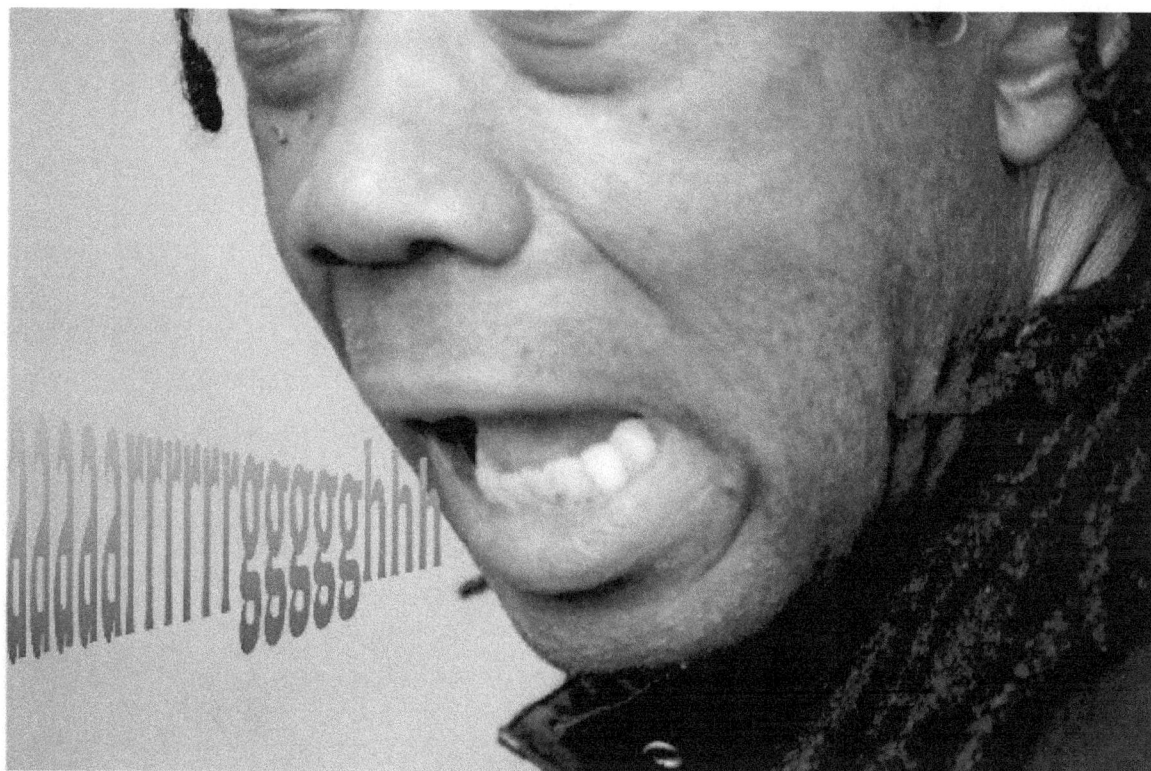

Dedicated to Conscious Ones....

Some of the content of this work included in *"(1 ~= ONE)" 2008

CONTENTS

...”{....

<u>FORMS/*”(contents)”</u>

Rubric

The UnderGround torrents of the Living Dead "Words" of Bishop Berkeley I visit at the Height of my Conscious Dreams....
It removes "I" to this page: a gaggle of neighbouring touristing ants watch in whispers of Unknowing Wonder as my Thoughts write these words....
The tourists go, and drive away: "Freedom, freedom!"
And the-It is aT 9.15 in the morning!

I the Awake, and "Rubicon" is the unknown Word sHe speaks....

"Guilty of *Loving* Yoo!" sHe Said...."*Leaving* Your Father's Way!"

A Beginning!

10:15 am 07/11/2007
Rainham, Kent

"0!":
<u>Storms</u>*faith*
<u>**Mythic Form contents**</u>

-a concrete re-Construction-
*(after
the risen sun:
ecOlogy myth,
Male-ism,
Egyptology,
Plato,
Pythagoras,
Hegel,
de Moivre,
Darwin,
Freud,
de Saussure,
Heisenstein,
Propp,
Levi-Strauss,
Lacan,
Barthes,
Stack Sullivan,
Williams,
Laing,
Derrida,
and, Feminism,
jaH-aYe:
the fallen night....)*

"0": Content forms mythics
The myth of the male beauty gOd aDonis

-a concrete re-Construction-
(after Raymond Williams)

0: Structures of: *"myth"*

mythics...
i.   *a risen sun:*
         summer noon

$F +(1/0) :\sim F +(iNfinity)$
     the River waxeth...

ecology myth
ii.:   *(after religion)*

$F (1/0) \sim "1" \sim F (oNe\text{-}iNfinity)$

Male
a: *after Male egO)*

$F -(-i. \ -i) :\sim F -(1/i)(1/i) :\sim F -(i^3)(i^3) :\sim F -(i^4)(i^2) :- a "1" = "too"...$

Hieroglyphics
b: *(after Egyptology)*

"0" : pTaH
"00": nU
"0!": ThOth
"1": Ra
"2": Osiris

an iNfinity of "one"...

and: "the lack"...
c: *(after Feminism)*

$F -(i. \ i) \sim F (i^2)(i^2) :\sim F (i^3)(i) :- a "1"...$

And: the thesis...
d: *(after jaH-aYe)*

Beauty male: female
         :~ contradiction : nexus
perhaps ! :~ iNfinity
            : divine

as gOd :~ male/female : male
-iNfinity : "1" : + iNfinity :~ too...

key –inFINITY := +InFINITY
         AS WHERE IS "1"?

IF "1" : (i^2)

Lim 1/0 :~ -iNfinity

Lim: 1/((1/iNfinity)--$\rightarrow$ Lim = 0) :~ +iNfinity

-iNfinity :~ "1" :~ +iNfinity

|0 1| :~ x = E [-1, 1] :~ |1 0 | := a "1" = a "2" ... "too"
|1 0| |0 1 |

Too, therefore... 1 : ½(2) : ½^(-1) + 2... "for"

Lim x = E [-1, 1] :~ |0 1| :~ | 1 0|
                     |1 0| |0 1|

½^(-1) + 1 :~ "too" : "for"...

Let us define -iNfinity := | 1 0|
                           |0 1|

therefore, an iNfinity of "one" := "2"

Let us complementarily define + iNfinity := |0 1|
                                            |1 0|

Proven, contradiction : beauty : | 1 0| :~ |0 1|
                                 |0 1| |1 0|

2 := 1
        is "too" = eternal One....

a fallen night, full-moon:
        winter midnight:

F (1/0) :~ F -(infinity)...
        the River waneth....

i.   **The parallels...**
ii.   **Venus....**

"00":   **iStory- the gift of "death"**

(Innocence "Before"
(i)
(he hid the nakedness from Him because he was "Aware")
(he kneweth "his nakedness" with Him because he was "conscious")

I.   The day...
II.   The night...
III.   The dawn...
IV.   The dusk, in a parallel world...
V.   aBraxas....

"Too":

Any the idea of a being relative, demands at least One The Being Absolute....

i.
<u>The parallels...</u>

i.

<u>Premise</u>   - the <u>fact</u> is sexy...

<u>Lie</u>   - theSis
           "The male knows nO beauty..."

<u>TRUTH</u>   - "Beauty" is, always loved....

<u>Anti-Lie</u>   - "Love" <u>is</u> <u>in</u> my eyes....

<u>Anti-Truth</u> - is the male, too
                              Sexy...

ii.
        "death" is
                always
                in
                        my eyes....
        "eyes" <u>are</u>
                in me.....

<u>a</u> male beauty, and
<u>b</u> the gift of "death"...

aDonis   - most beautiful gOd: best loved
Zeus   - prophecy aDonis will die...
Zeus   - asked all things to promise never to harm the gOd; except the thorn.

Thorn kills aDonis...
He goes to the Underworld/Mourning of world....

a   an

"1" ~ +iNfinity

beauty...

> I, this
>> Alowd, living:
>
> Live with that
>> Known, awakening:
>> Awake in You
>> Loving the awake…

b   the

"1" ~ -(iNfinity)

"death"....

> and, that
>> belittled, death:
>
> die in this:
>> unbeknown, sleeping:
>> asleep to Thou
>> hating being,
>>> asleep….

<u>a</u> sun, and
<u>b</u> a skin-tone...

How the sun
Made
Black into White...

<u>a</u>   <u>the-sis : beyond the risen heat of the sun...</u>

<u>sisters...</u>

F -(i. i) ~ F (i^2)(i^2) :~ F (i^3)(i) :- a "1"...

skin tone : sun heat

<u>brothers...</u>

F -(-i. -i) :~ F -(1/i)(1/i) :~ F -(i^3)(i^3) :~ F -(i^4)(i^2) :- a "1" = "too"...

skin-tone : sun heat

<u>b</u>   <u>doings</u>

the temperature... falls/rises...

she... +/-   a, living with:
              the sun
         out, knowing, doubting another:
         fat, pregnant of the sun/brr
              of mothers
              of night
         fathers of night and day

he... +/-   an asleep, within:
              the skin of night
         in unknown, faith:
         muscled, fit to hunt/gatherer
         father of sons, of night/day
              daughters of day and night
         high temperature:
              dark lovely of skin-tone
              as follows sun,
              summer has,

as a lightening into dark;
protection from
      the can-saws of light;
waker on a wet day
black his colour, perfect
perfected learning of all white...

<u>a</u> meal, and
<u>b</u> "eat"...

Of culture....

<u>a</u>   <u>the-sis : beyond the meal...</u>

<u>sisters</u>...

F -(1/0) :~ F +(iNfinity)...

food : culture

<u>brothers</u>...

F (1/0) :~ F -(iNfinity)

food : culture

<u>b</u>   <u>doings</u>

the burden... falls/rises...

she... +/-   (a,
                difference, being
                    aNother's
                sleep gathering for
                    the difference:
                aNother's feeding
                    different being
                biG to risen, burden being
                biG to different
                    Being,
                    Being being....)

he... +/-   the,
                similar, being
                    one-aNother's
                sleep hunting for
                    the similar:
                one-aNother's feeding
                    similar being
                biGGer to risen, burden being
                biGGer to similar/
                    Being,
                    Being being....

<u>a</u> sex, and
<u>b</u> "life"...

Sexuality....

<u>a   the-sis : beyond the sex...</u>

<u>sisters</u>...

F (i.i) ~ F (i^2)(i^2) :~ F (i^3)(i) :- a "1"...
F -(1/0) :~ F +(iNfinity)...

sex : life

<u>brothers</u>...

F -(-i. -i) :~ F -(1/i)(1/i) :~ F -(i^3)(i^3) :~ F -(i^4)(i^2) :- a "1" = "too"...
F (1/0) :~ F -(iNfinity)...

sex : life

<u>b   doings</u>

the passion... falls/rises...

she... +/-   (pleasure, being
                suffering, being
                being passion:
        love lust an excellence....)

he... +/-   lustful, being
                bliss, being
                being passion:
        love lust a bit of all right....

          <u>a</u> faith, and
            <u>b</u> "survival"...

Religion....

<u>a   the-sis : beyond the faith...</u>

<u>sisters and brothers</u>...

F -(1/0) :~ F +(iNfinity)...

faith : survival

<u>b   doings</u>

consciousness... falls/rises...

sHe... +/-   knowLedge
              an Excitement
              rises beyond
                  the exCitement ledge
              being beyond the ledge of
                          exCitement to
              be-In....

"0!":

<u>Storms*faith*</u>
<u>**Mythic Form contents**</u>

-a concrete re-Construction-
-transformations-
(after
"nO",
"The fallen night",
"religion",
"ecology myth",
"Egyptology",
Plato,
Pythagoras,
Descartes,
Hegel,
de Moivre,
Darwin,
Marx,
Freud,
Adler,
de Saussure,
Heisenstein,
Propp,
Jung,
Levi-Strauss,
Lacan,
Barthes,
Stack Sullivan,
Williams,
Gramsci,
Sartre,
Laing,
Foucault,
Derrida,
pTaH-aYe
"the risen sun",
"aYe"....)

"0":   <u>**Content forms mythics**</u>

-a concrete re-Construction-
(after Raymond Williams)

Air – You enslave me to nO-thing, sinO-Marxist literary tradition….

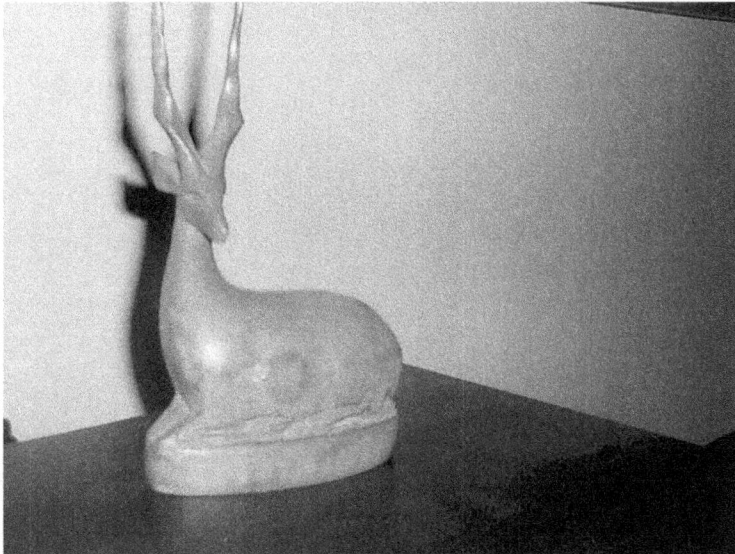

0: Structures of: "myth"

mythics nO… (i^2)

(after "death")

The risen sun,
Summer noon…
The River waxeth…

Lim |1/(1/iN-finity)/"0" | ----→ rising +iN-finity…

ecology myth

one: (after religion)

Lim (1/0) :~ "1" :~ F (oNe-iNfinity)

Hieroglyphics

too: (after Egyptology)

"0" : Ptah

"00": nU

"0!": ThOth

"1": Ra

"2": Osiris

"3": pharaoh

an iNfinity of "one"…

and: "translations"…
a: (after Pythagoras)

(a^2) + (b^2) = (c^2)

or,

root (1/(x^2) + 1/(y^2)) = 2((x^2) + (y^2))

or,

root Y = 2(-Y) *or,* ½ ~= (i^2)

24

and: **"essentials"...**
b: *(after Plato)*

$$F_{(i^2)} \sim F_{((-1/(i^2))}$$

and: "too" myth...

c: (after Hegel)

("0": (i^2)):Thesis :~ ((i^2): 1): anti-Thesis :~ (1:1/(i^2):2): Synthesis

and: mind/body dualism

d.:   '   *(after Descartes)*

(1 : (i^2)) :~ "2"

and: "natural waveforms"...

e: (after de Moivre)

$$F_{(e^{\wedge}(i.pi))} \sim F_{-(e^{\wedge}((i^3)(pi)))}$$

and: "genetics"...

f: (after Darwin)

Lim F -(1/0) :~ F +(infinity/infinity) :- F +((infinity + 1)/infinity) ~ F +(1/0)

and: "anxiety"/"security"

g: (after Kierkegaard)

Lim (1/iN-finity) ---→ :~ (1 . "0")

and: the "materialist dialectic"

h: *(after Marx/Engels)*

now

-time (quantity)← || (I^2) : "1" :~ | 1 0 0 || = E (+1, +iN-finity)→+time quality)

| 0 1 0 |

| 0 0 1 |

and: "preScient" myth....

i: (after Freud)

|1 0 | ~ | 0 1| ~ | 1 0 | ~ | 0 1| :~ *F (female)* :~ *F (male)* | 0 1| |1 0 |

:~ F (i^4) :~ F -(i^3)(i^2)

and: the will to power

j: after Adler

1 : (i^3) ----→ iN-Finity….

and: "signs"...

k: (after de Saussure)

Sr : F ("0") ~ Sd : F -(1/"0")

and: "the observer interferes"...
l:(after Heisenstein/Freud/Adler/Jung/Stack Sullivan/Lacan/Sartre/Laing/Derrida)

$$F \text{ -(i^4)(i^2)} : F \text{ ((i^3)(i^2))(1/((i^3)(i^2)))}$$

and: "for" myth...

m: (after Vladimir Propp)

"0":   0 : Equilibrium
(i^2): 1 : dis-Equilibrium
-(i^2): 2 : Search
(i^3): 3 : re-Equilibrium
(i^4): 4 : Celebration

and: the anima

n: (after Jung)

F (male) ~ F (i^3)(i^2) : F (i^2)

and: "naScent" Myth...

o: (after Levi-Strauss)

The function of one, according to the function of another, is in a relationship with The function of the one, according to the function of the other, <u>translates to</u>

The function of one, too, according to the function of another, too, is in a relationship with The function of the one, too, according to the function of the other too…..

D: *F (male)* $= F \quad ((i^3) : (i/2)) : F \quad (2 : 1 : \frac{1}{2})$

$$\underline{(1/2 : 1 : 2)} \qquad ((i^4) : (i/2))$$

$\sim e = F \,(i^4) : F \,(-1)(i^2) \quad (i^4)$

or,

$e = F \quad ((pi)^{bn}) : F \quad (-((pi)^{bn}) \sim f = F \quad ((i)\,^{4an}) : F \quad (i^2)((i)\,^{4an}))$

$\underline{(I^2)(((i)\,^{4an}))} \quad \underline{a((i)\,^{4an})} \qquad \underline{((pi)^{bn}} \qquad \underline{((i^2)((i)^4) \,.\, i)((pi)^{bn})}$

and: "neUn consciousness"...

p: *(after Lacan)*

"0" ~   **"00"** ~   **"0!"** :   **+/-(------->) ~ "1"**
sleep ~ **waken** ~   **awareness :   a rise ~ enlightenment**

<u>and: "rhetoric"</u>...
q: *(after Barthes)*

pre-premise, $-(i^2)$: thesis, (i^2) : hypothesis, (i^4) : premise, $4(i^2)$: re-premise, "1"...

<u>and: "the uncanny"</u>...
r: *(after Stack Sullivan)*

$F _{-(i^4)(i^2)} \sim F _{(i^3)(i^3)}$

<u>and: "Form/content"</u>...
     s: (after Williams)

$(i^2)(1/(i^2)) : +(i^4)$

<u>and: anxiety</u>
t: *(after Sartre)*

$(I^3)(i^3) \sim (i^2)$

excited → stressful → traumatic   → dangerous →   deadly

disquiet → anxious   → nervous   → fearful   →   frightful

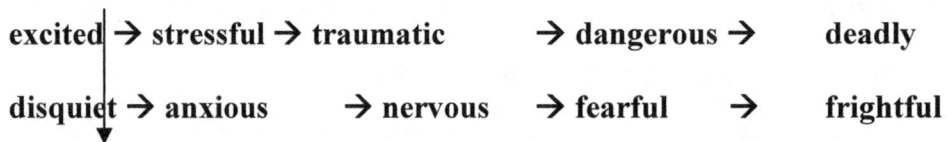

<u>and: "ontological insecurity"</u>
u: *(after Laing)*

$F _{-(i^2)(i^4)} \sim F _{(i^3)}$

and: "difference"...
v: (after Derrida)

"1"   ~   **"2"**
-(i^2) ~   (-1/(i^2))

---

and: SUMMARY - the thesis

t: (after pTaH-aYe)

i.

if origin of the real line: "0"

and, x = E [-1, 1]

this closed interval ~= "2"... "too"

if -1 :~ i^2 = positive real...

then, | i^2 | :~ ½ ...

and, -1----→ -iNfinity :~ (i^2) -----→ "0"....

ii.

Beauty male: female

$: F (male) = F \frac{((i) : (I^3)) : F (2 : 1 : ½)}{(1/2 : 1 : 2) \quad ((i) : (I^3))}$

$F (FeMale) = F \frac{((i^2) : (I^2)) : F (2 : 1 : ½)}{(1/2 : 1 : 2) \quad ((i^2) : (I^2))}$

:~ contradiction : nexus
perhaps 1 :~ iNfinity

: divine
as gOd :~ male/female : male

-iNfinity : "1" : + iNfinity :~ too…

key –inFINITY := +InFINITY

       AS WHERE IS "1"?
       IF "1" : (i^2)

Lim 1/0 :~ -iNfinity : Male…

       Lim: 1/((1/iNfinity)--→ Lim = 1) :~ +iNfinity : "feMale"

    -iNfinity :~ "1" :~ +iNfinity

more,

$$|1\ 0| \ :\sim\ x = E\ [-1,\ 1]\ :\sim\ |0\ 1| := a\ \text{"1"} = a\ \text{"2"}\ \ldots\ \text{"too"}$$

Too, therefore… 1 : (½(2)) : ((½^(-1)) + 1)… "for"

$$\text{Lim } x = E\ [1,\ 2]\ :\sim\ \begin{matrix}|1\ 0| :\sim |0\ 1|\\ |0\ 1| \quad |1\ 0|\end{matrix}$$

((½^(-1)) + 1) :~ "too" : "for"…

:~ |1 0 0|
          |01 0|
   |0 0 1|

       therefore, a "one" := "3"

      Contradiction : beauty : | 1 0| :~ | 0 1|
                            | 0 1| |1 0|

"1" := 2

       is "too" ~= eternal One… "Male"…

       the feMale:

       (i^2) ~ (i^3)(*i*^3)

          dialectic

       feMale ~ fe-Male

iii.

"3" is a myth…

the referents of "0"… 1-2: *too*, provides "3"…

3 is/not a prime number: is a *too*…

Logos *too - (includes idea of "4")*
"0": premise - "1": thesis - "2": anti-thesis - "3": synthesis

iv.

"0"… 1-2-3-4 ------→ "5" is a base fractal in Nature…

…And of the nature of ideas….

v.

The Universe appears to be an 4-dimensional time-space continuum, with eight referents to time in space….

out-

a fallen night,

winter midnight…

the River waneth…

Lim |1/1/(1/"0") | ----→ falling -iN-finity….

mythics aye…

| 1 0 | ----→ E (the real)
| 0 1 |

(after "Life")….

"too" —————————> to "00"

"0"   Grammar….

"00":   doings of myth to "drama"….

0: Grammar

{/	**Start, Or of PROGRAM…**
"0"	**the first utterance of nothingness…**
"0!"	**the first utterance of infinity…**
"0"	**an utterance of nothingness…**
"0!"	**one utterance of (an) infinity…**
,	**refers…**
'	**defers…**
;	**referral/deferral…**
"-"	**a mythics…**
:	**relates to…**
:-	**similar to…**
( )	**function of…**
(--)	**usually, a function of the future-past (a tomorrow)…**
+	plus/positive…
-	minus/negative…
E	**sum of elements of set…**
| |	**determinant : (integral value of transposes to equal nothing)…**
!	**a transport into being (infinite)/probability…**
~	**transforms into…**
:~	**transposes into…**
:=	**relates to and equals…**
~=	**~ :~ translates to equal…**
=	**equals…**
---->	**slide to: a "do", now to future…**
<---	**slide back to: often a "do" now to past…**
.	**ends/multiplies: an "act"…**
...	**produces a catastrophic-like ending: a "work"…**
....	**produces a fractal-like "ending": a "happening"…**
.....	**PRODUCES A MYTHIC CONCLUSION: a "demonstration"…..**
___	**GOD-ly…..**
" "	**an utterance, a spoken with/of, a <u>making-the-being</u>……**
a/A	**some-only/"one" of…**
&	**and" : plus, a *too*…**
the	**certain-one…**
eLSe	**…other/or….**

*   context....

i/I   (1/(root 2))/(special referent to a root of one-too)...

too "mythic one-and"...

premise   beginnings
        : | "0!" |...

hypothesis next essence of new idea after a beginnings
        : ("0!") . (i^2)...

thesis   next proposition after a beginnings
        : ((i^2) & "1")...

antithesis next contradiction to thesis
        : ((i^2) & "2")...

synthesis   next contradiction to antithesis
        : ((i^2) & "3")...

condition   a necessity that...

do   discovery...

act   proof....

play   discovery by analysis...

demo   demonstration....

real   fact in Nature....

irreal   fiction/irrationality/absurdity in Nature....

reAL   controversy in "it"/text....

nULL!   an a/the "absurd/illogic" in the nature of the Universe....

nO!   an a/The "illogic" in the nature of the Universe....

yeah! : perhaps an &/the_Absurd in the Nature of the Universe.....

yes!   an/the "Logic" in the Nature of the Universe....

aye!   an/The_"Logic" in the Nature of the Universe....

*"((bLACK_bOX) : (WHITE_bOx) : (bLONDe_bOX))" :
   :*"(The "Integrated_WAYS" in the
   Nature of the space-TIME_ContinUUms)"....

The fLUX   The eBB/Flow of ALL in Nature....

T**O**T   *"(aBraXaS NATURE)"....

Ga-GA   The Way of Chaos/Catastrophe in nAture....

God/gOd/God/GOD rising forms of godliness in Nature....

IottOi   The_One-GOD in "T**O**T-Nature"....

....enDs}ENDS....

End of PROGRAM….

…."{….

"0": Form contents mythics
(after Raymond Williams)

1: Content forms mythics
(after Vladimir Propp)

0 : Equilibrium
1 : dis-Equilibrium
2 : Search
3 : re-Equilibrium
4 : Wedding

(Innocence "Before"
(i)
(he hid the nakedness from Him because he was "Aware")
(he kneweth "his nakedness" with Him because he was "conscious")
:~   ("I")

1:   Content forms mythics

(after
Sigmund Freud
Ferdinand de Saussure
Vladimir Propp
Claude Levi-Strauss
Noam Chomsky
Roland Barthes
Raymond Williams
Jacques Lacan
Jacques Derrida)

"1":   neUn maths

0.   *(Theses)* Premises - Doings

1.   "0"   ~   (1 x "0") = utterance of nothingness = a real : God's Word

2.   "0"   :   "1"   ~ = 1 &... = a real

3.   1 &...   :   "2"   ~ = 2 &... = a real

I.   *(Anti-theses)* Theses - Actions

1.   "0"   ~   utterance of nothingness = real ~ WORD ~ Thoth/Nu ~ God

2.   "i"   ~   fourth root of 1 = real

3.   "i^2"   ~   root 1 ~ = -1 = real

4.   "i^3"   ~   -i   ~   fourth root of 1 = real

5.   "i^4"   ~   "1"   ~ =   1 = real

6.   "pi"   ~   3 + j = approx. 3.1415926 = real

7.   "1"   ~ =   1^0   =   1   = real

8.   1 + 1   ~ =   2^1   =   2 = real

9.   1 x 1   ~ =   2^0   ~ =   2   =   1 = real

II.   *(Syntheses - Dialectics)* "Lies" - Works/plays

1.   "0"   ~   approx. 0.83

2.   "i"   ~   1/(root 2)

3.   "i^2"   ~   1/2

4.   "i^3"   ~   1/2(root 2)

5.   "i^4"   ~   "1"   =   1

6..   pi   ~   3 + j = approx. 3.1415926

7.   "1"   ~   "2"   ~ = 2

8./9.   (1 + 1)/(1 x 1) : (1 x 1)(1 x 1) : & ~ "2" : *Too* ~ = 2 ~ = 8+ : *river* ~ = 4 : 4 :

                              ...for

Lies to sexuality:

M   : male
f   : femme
F   : woman

Male	boy	male-femme	girl	woman
M:	M/f:	M/F:	f:	F:

Sexual Types

Asexual   bi-sexual gay gaya

IV. (*Demo*) "Myths" - Demonstrations

Content "for" Mythics
(after Claude Levi-Strauss:Greek myths are structures of how to form "2" from "1")

(f-o-r = from/"f-o-r-m" = dark sign "o" present/move)

i.

Strauss' *Greek Myth*s F_a (x) : F_b (y) ~ F_a (y): F_{-x} (B)

Reappraisal Greek Myths: $A = F$ (x) : F (y) ~ $B = F$ (b): F (Y)

$\underline{a = -(i^2):i^4}$ $\underline{b = i^2:-(i^4)}$ $\underline{x}$ $\underline{a-1}$

Translates....   **1/-1:   -1/1   ~-(1+1/1x1): -2/2:   "0"/"0"**

Which translates: "Functions of one and its contradiction, is *transformed* into two, contradicting the Word which is God", eg., contradicting functions Y = -x ; x = -Y......

A man and a woman contradict and are *transformed* into a union of *too* **(light or dark)**, in a contradiction with the forces of the WORD which is God...*For example, Adam and Eve in Eden...*

ii.

*"Too" Myth*s: $A = F$ (x) : F (y) ~ **"Too"** $= F_{}$ $\underline{(i^2}:-(i^4):}$ F (Y)/"1"

$\underline{-(i^2):i^4}$ $\underline{i^2:-(i^4)}$ $\underline{x}$ $\underline{-x/i^2}$

Which translates: "Functions of One and its contradiction, is *transformed* into two contradicts to it = Too", eg., contradicting functions Y = x + 1; x = Y-1, which yields:

1/-1   :   "0"/"0"   ~   x^2:$\underline{i^2/1^2:1+1/1x1}$:-2:   -2

A good man contradict with the forces of the WORD and are *transformed* into a union of *too* in a contradiction with another union of *too*... *For example in the Bible when chosen man Adam forms a union with Eve, and their too is contradicted by the too of the field...*

iii.

*iOj Myth*s: $\mathbf{C} = F$ (B = (x^3 + 1)) : F (y^3 -1) ~ **River** $= \mathbf{D} = F$ (i^3): F (y^3 - 1)

$\underline{- i^3}$ $\underline{i^3}$ $\underline{(x^3 + 1)}$ $\underline{i^3}$

Which translates: "Functions of Too and its contradiction, is *transformed* into a discourse of Eight, in a contradicts to the WORD ", eg., contradicting functions :
Y = x^3 + 1; x = Y^3-1, which yields:

"Too" = -2:-B = 2   ~   **River** = 6+2i:**8+**:   i^3 = i^2 = i^3(1-(i^4)) = **"0"**

A union of *too* in a contradiction with another union of *too*... are *transformed* into the relay of a discoursive fractal of eight, which contradict the WORD...The Too of Adam and Eve, yielded a discoursive fractal, or *River*, in later tales of Noah, Abraham, Lot, Moses, David, Solomon, Jesus Christ, and Mohammed, and Yogi Singh....

iv.

*one-Self Myth*s: ~ **River = D** = F (i^3) : F (2 : 1 : 1/2) ~ **e** = F **(i^4) : F (-1)**
                          (1/2 : 1 : 2)   i^3   **(i^2)   (i^4)**

Which translates: "Functions of one-Self and its contradiction, is *transformed* into a discourse of 4 = -4, in a contradicts to the WORD ", eg., contradicting functions:
                          Y = 2/x = 1/2x; xY = 2 = 2Y: which yields:

~   **River** = 6+2i:**8+**:   i^3 = i^2 = i^3(1-(i^4)) = **"0"** ~ **1/2/4:4:8 : (1/2)/1:-1**

A union of *too-River* in a contradiction with another union of *too*... are *transformed* into the relay of a discoursive fractal of four = "*for*", which contradict the WORD... The Too of Adam and Eve, yielded a discoursive fractal, or *River*, in later tales of Noah, Abraham, Lot, Moses, David, Solomon, Jesus Christ, and Mohammed, and Yogi Singh... yielding a modern myth of the *neUn*:

for : four ~ one : body-trinity : one body/consciousness/spirit/soul...
...neUn ~ one body/consciousness/spirit/soul : one the
real/sure/certain/concrete....

Form/content mythics
(after Ferdinand de Saussure)

```
Signifiers    -      "0"------------> 1: "Too" : "for"
                  4 : 5!              /
                  3<--------------2
```

Form content mythics
(after Roland Barthes)

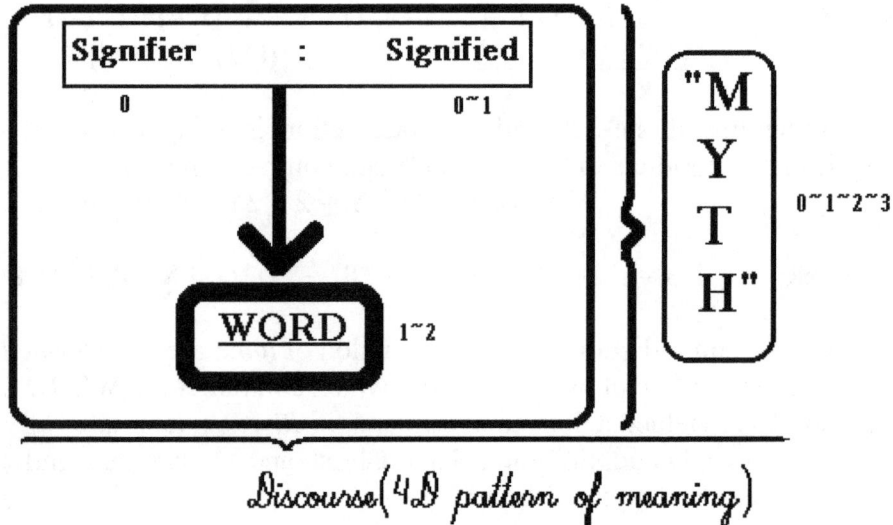

Discourse (4D pattern of meaning)

Myth

"0":1-->A *Signifier* refers does at/to the floating utterance of nothingness ("0")...

1:2---> A Signifier always acts as a defers to itself of a *SIGNIFIED*...

2:3---> A *WORD* makes a referral/deferral to A WORK/PLAY OF MEANING, a "lie"...

3:4:   The PAROLE of word meanings is relayed (referral/deferral!) in the total language as a "lie-line", a "path to truth", a happening...

4:5:.. Each lie-line is a "MYTH", and owns a pattern of speech, a "DEMO", which is its mythic "truth"...

V. (*Tiers*) **Mythic Discourses - Events**

5:6...   PROGRAMMING:

Signifiers -   "0"----------------------> 1: "Too" : "for"
                       4: 5!:6!   /
                       3<-----------------------2

Each "DISCOURSE" owns an program of events, a " TIER", which is its "real" rhetoric.

VI. (*Catastrophes*) **"Sect" - Drama**

6:7. CATALOGUE:

Signifiers -   "0"------------> 1: "Too" : "for"
                        4: 5!:6!:7!   /
                        3<--------------2

Each "SECT" owns a catalogue of drama, a "CATASTROPHE", which is its "sure" need.

VII. (*Fractals*) **"Cult" - Episodes**

7:8: LIBRARY:

Signifiers -   "0"----------------------> 1: "Too" : "for"
                        4: 5!:6!:7!:8!   /
                        3<-----------------------2

Each "CULT" owns a library of episodes, an EPIC, which is its "certain" greed.

VIII. (*Chaos*) **"Faith" - Incidents**

8:8+: *WEB*....

Signifiers -   "0"----------------------> 1: "Too" : "for"
                        4: 5!:6!:7!:8!:8+!   /
                        3<-----------------------2

Each "FAITH" owns a web of incidents, a SAGA, which is its concrete mood.

2.   Instincts

No genes: *("Lies" - Paths to Truths)*

Form No Categories   *Form Mythic No Categories.................*

IDEAS - Selfish Reader Amino-Acids
(after Jacques Lacan)

--------------------Structured 4D------------------						--------------------Structured 8D---		
--------"0" ~1---	--- 1 ~ 2--	---2 ~ 3--	---3 ~ 4--		---4 ~ 5--	---5 ~ 6 --	---6 ~ 7 --	---7 ~ 8--
"0" ~ 1	1 ~ 2	2 ~ 3	3 ~ 4	4 ~ 5	5 ~ 6	6 ~ 7	7 ~ 8	8 ~ 8+ = 9
refers defers	referral defferal	feed	dial	exit	rise	drift	have	haven
refers defers	protect secrete	guess connect link	know	conscious	enlighten	become	ascend	transcend
signifier no	signified no-no	word aie	parole aye	speech yea	rhetoric yes	need sure	greed certainly	mood concretely
un-truth thesis	lay anti-thesis	lie synthesis	truth happening	myth demo	discourse tier	sect catastrophe	cult fractal	faith chaos
impossible per-	possible haps	probable per-haps	true happening	myth demo	real event	sure drama	certain epic	concrete saga
no-time doing	now action	past work/play	now-past happening	now-past-future demo	now-future-past event	forever-now drama	eternal-now episode	eternal-forever-now incident
black abyss	red sun	orange star	yellow sky	green heaven	blue universe	indigo abraxas	violet diety	psychedelic gOd
instincts anal	body oral	mind sexual	spirit spiritual	soul soul	real the-real	sure the-sure	certain the-certain	concrete the-concrete
asleep refering- -defering	awake protecting- -secreting	aware thinking- -connecting- -linking	knowing dream	conscious demonstrate	enlightened evene	becoming drama	ascending episode	transcending create
refer defer pleasure	protect secrete satisfaction	love come	adore know	possess own	contemplate bliss	haunt obsess	have behave	haven blythe
deny mortal	cause living	pretend uttering	tender dead	conceive living-dead	receive deading-dead	bind live	pretext live-dead	text immortal
that	what	that-it	that-you	that-You	that-we	that-they	that-one	that, the
who	I	I-thou	I-it	I-We	I-You	I-They	One	I-The
the	my	mine	me	I	myself	the-one	I, a	I, the

Machine Coding - Selfless Writer RNA
(after Noam Chomsky)
Deep-structured language

O	Oo	a	ah	awe	*a-awe*	*oo-awe*	*oo-awe*	*awe-awe*
a	ha	aha	aha-ah	aha-aha	*a-a-aha*	*a-a-a-aha*	*a-a-a-a-aha*	
e'	ee	ee-ee	e-e-e	e-e-e-e	*e-e-e-e-e*	*e-e-e-e-e-e*	*e-e-e-e-e-e-e*	
e'	eh	eh-ee	eh-eh	awe	*a-awe*	*oo-awe*	*oo-awe*	*awe-awe*
ma	nana	sah	baas	ra	*ra-rah*	*ra-ra-ra*	*ga*	*gah*
fi-mi	fi-yu	fi-s/he/it	fi-wi	fi-you	*fi-dem*	*fi-oonu*	*fi-awe*	*fi-gah*

GENES   -   Social Programming DNA
(Social discourses, human dimensions)

perfect	same	similar	identical	special	*species*	*environment*	*system*	*ecology*
imperfection	flawed	dis-similar	different	mutation	*strange*	*alien*	*foreign*	*outside*
peace	hearth	home	haven	temple	*cathedral*	*tomb*	*sepulchre*	*heaven*
crisis	fight	conflict	battle	war	*conflagation*	*holocaust*	*apocalypse*	*era*
instincts	body	mind	spirit	soul	*real*	*sure*	*certain*	*concrete*
individual	couple	family	tribe	nation	*people*	*world*	*universe*	*heaven*
I	thou	s/he/it	us	ours	*people's*	*world's*	*Ra's*	*God's*
s/he/it/one	couple	family	tribe	nation	*people*	*world*	*universe*	*heaven*
s/he/it	us	ours	We	You	*They*	*The-It*	*Ra*	*God*
son	father	grand	great	nana	*nenny*	*ninny*	*nonny*	*nunny*
daughter	mother	grand	great	nana	*nenny*	*ninny*	*nonny*	*mumy*
son	father	grand	great	nana	*nenny*	*saint*	*Christ/Buddha/Ram*	*angel*
son	mother	grand	great	nana	*nenny*	*Madonna*	*Mary/Maya/Sita*	*angel*

Go! Translator

a Oo	=	I am a powerful being
aha	=	I understand
nana a ra	=	My grandmother is a queen
a fi-wi baas	=	This is our ruler
e' eh-eh	=	He/She's actively acting like a high god
e' awe-awe	=	He/She's the biggest type of person

3. Drives to consciousness
(after Sigmund Freud)

Being lusts in "nothingness": a kept vacancy at the roots of "sleep": "*0*":..

One's primary spring is to grow:

contradictions, "i", a drive to to "survive",
an instinct being a to eat and a to drink, being
    a waste to be anal and urinal:
        considerations of "wasting" being a driving instinct:
            a cry, from a drive often to make/too create from,

contradictions to i, being genital, instincts of sex: to reproduce:
    a desire from need, a drive to create anew,

two both too, both "too" to: creating "I", being : moods of being:
    abouyance: "spurring" you to crisis
    contradicting, *evvayance*: "lacking", making you stop from doing:
"death" instinct:
    deferring to *joyssance*: "blissing", encouraging you to doing,
    contradicting *annewsance*: "pain", stopping doing:
"life" instinct:

all process of I/thou: neGus -sleeping, a kept/vacancy at the start of consciousness
                  neAus -awakening, insight that spurts creation
                  neXus -awareness: thinking, research, translation
                  neYin -ennui, a loss at the feeling of the new/rising
                  neHus -"enlightenment", seeing "anew"...

Lusts being considerations of **bliss**:
                "*0*": *abouyance*: *evvayance*: *joyssance*: *annewsance*:
    anal, a passive response to "feeding", being *oral*
    oral, an active response to "genitality": being *sexual*

passive/active responses:
    sexual, comforting reassurance, a response to "loving", the *spiritual*
    spiritual: tempting/tormenting, a response to "adoring", the *soul*
    soul: possession, a response to "haunting", the *real*
    real: obsession, a response to "contemplating", the *sure*
    sure: blessing, a response to "saving/serving", the *certain*
    certain: salvationing, a response to "holy-ing", the *concrete*...

this being "holiness" : *lust satisfaction* being a **blithe**: a programming making a
web....

4. neUn consciousness
(synaptical responses)

"0": neGus :   passivity : stasis : *sleeping* : kept *vacancy* at a beginning...
                                    -------> *refers* - aways
                                    -------> *defers* - allows

"i": neAus :   activity : conflict : *awakening* : *insights* starting creation...
                                    ------> *protecting* - allows to
                                    ------> *secreting* - allows from

"i^2": neXus : choice : problem : *awareness* ----> thinking *: guessing*
                                    connecting *: researching*
                                    linking *: translating*

neYin :   the loss at *rising to* <u>One</u> leading to *ennuie*,
          feeling of "new"/boredom ------> a "glip"...

<u>1:</u>   <u>neHus</u> : (first *too* quanta) : myth-ing : *lie-ing* : knowing

<u>2:</u>   <u>neGus-neHus</u> : (second *too* quanta) : diffrancing : *conscious* : discoursing

<u>3.</u>   <u>neAus-neHus</u> : (third *too* quanta) : realising : *enlightened* : programming

<u>4.</u>   <u>neXus-neHus</u> : (fourth *too* quanta) : structuring : *sure* : cataloguing

<u>5.</u>   <u>neHus-neHus</u> : (fifth *too* quanta) : making true : *certain* : totalising

<u>6.</u>   <u>neGus-neHus-neHus</u> : (sixth *too* quanta) : constructing : *concrete* :
                                    systematising....

{...5.　Go! translating Mythic Alphabet
(after Egyptian heiroglyphics)

a	-	sleeping one/thing/"1"
b	-	be/big
c	-	sighted
d	-	do
e	-	awake one/the spirit in the thing/he, she, it
ee	-	*aware one/consciousness/"I"*
f	-	force/sign dark
g	-	sign light/vehicle
h	-	home/heaven
i	-	aware one/consciousness/"I"
j	-	just/joy
k	-	know
l	-	here
m	-	move
n	-	negatively top/dark chief/black ruler
o	-	connecting one/matter/"0"
oo	-	*linking one/person/"you"*
p	-	take it/leave
q	-	give it/stay
r	-	presently/are
s	-	understanding
t	-	the law/the way/money
u	-	linking one/person/"you"
v	-	two/too
w	-	route
x	-	connecting/suffering
y	-	knowing one/soul/"why"
z	-	end

_____

Letters joining-up add "to" in their middle:

e.g., "oo" = "connecting one to connecting one"

= "**one** connecting one"

= "linking one".

Letters join-up to make words,

e.g., "book" = "be to linking one to know"

= "be you know"

"The Book of the Dead" means:

"the way e be you know connecting one to force the Way e a do"....

Knowledge

a	e	I	O	u	y	"iou"	"iously"	sure
dreamt	*protected secreted*	*thought/aware*	*connected*	*linked*	*known*	*conscious*	*enlightened*	*become*
sleeping	waking	aware	aware	aware	knowing	conscious	enlightening	become

...}....

6.   one-Self

(after Standard English)

Word knows excellence of *no*-thing in all *things, body, one*: the tribute of the River;

Something, Somebody, Someone holds the keys:
Anything, Anybody, Anyone can the further stairs, and scent the Heaven sent;
Everyting, Everybody, Everyone has the dream:
All-of-one: body: thing knows

gOd: *God* is almighty;

Here a *Person body* speaks

To *Child*: is

Unborn to

The *Spirit* awakening...

Love is the inner winding chords to *Too*: *soul*:

and, Love is *for*: the *Real, Sure, Certain: Concrete*....

7.   Intro to duB-poetry

The term *duB-poetry*, meaning "poetry with a musical rhythm", usually reggae rhythms, was coined in the 1970's by duB-poet. Oku Onoura; the genre had already been explored by ohn Cooper Clarke, and Linton Kwesi Johnson in Britain and others in Jamaica.

Some of the roots of duB-poetry lies in the Jamaican dj's like I-Roy and Big Youth, and comes from a tradition of poetry set to music harking back to the Last Poets, in the United States, and Beat Poets in the 1950's and 1960's. As a form, duB-poetry is a survival through slavery of the African praise poets, the *Griots*, who used songs to their kings and others in the culture, and acted as folk historians.

As praise poets, the duB-poets were journalistic spokespersons for the common folk, and spoke up about news events in the history of the culture they lived in, giving praise or curse accordingly.

Nowadays, a derivative of the griot-style, rap, is commonplace, but most commonly, with the political stance not always there, the rappers boasting, or talking grandly of themselves...

Dub-poetry dealing with the politics of personal relationships is not too common, and is the essential feature of T-site. I have taken some of the rhythms too, away from much of their traditional bases and into the classical genre...

Sexuality/sensuality is one of the themes of T-site duB-poetry, and one of links and lens to understanding in an attempt to show how I believe it to be central in all our lives...

The poetry/music is a bare allegory of several world myth, a text based on the narrative of another text. The form/content is "mythic", following my theory of mathematics, especially regarding mythic discourses...

The images in the text are complex, especially as I don't always use simple metaphors, or anaematopia, but often construct a phrase or sentence so the meanings 'explode' in clusters and intentionally there must be several interpretations. This is a latter day influence on my poetry coming from French aesthetic theory. I do this to allow the reader to *write* meanings into the text; also have a choice of meaning, so that I as writer, and she/he are jointly constructing relationships: that is, *working* with words/music to produce a text. This is analogous to an element of free will...

Furthermore, no longer influenced by the Imagists and Hans Magnus Enzensberger, I now believe that images must be *startling*! and a worthwhile text must be struggled with, and enjoyed again and again; as well as being 'written' too by the reader.

Thus, a lot of my images exists as French critic Roland Barthes explains, as *elisons* : meanings that you can <u>just</u> understand as a reader, and when <u>held</u> in the mind, they 'slip' and you have to work to make them plain...

Too, there is a construct around a time base of eternity: the eternal-forever-now....

….”{….

….intro

“that”

((the)_MiMe))

egO in Self; power the bAse of sexuality:

$$\text{power} \sim: \text{sexuality} \sim: \text{"I"}\dots.$$

"I!"…
"I" Am….
"I" refer only to <u>ME</u>!

"You!"…
"You" are always referring back to "I"-through-<u>Me</u>!….

"I!" Am I to "I"….: "One!"….Alone….
"You!" are you! to "You"….: "Two!"….

i

"this"

1….

"I" to suffer You!
….And I not to suffer!
To suffer <u>Me</u> is wrong!

There:
  Being A God!

…Be_Cause:
   I sleeping in High_Places….

Defence_Only-in-The_Nature, Child….
Attack_You-to-Peace!

….Sleep-Always-in-The_Awake!

ReAl!….

How little sHe questions You to this!

2....

Compliment to Spring the rising wood....
        A HeartBeat the Seed!
A Gathers as She Questions The_Heart....
        Coo_Meant_on, The_Wind....

...I Mean You as "I" don't Suffer_To....
Suffering_Me You die!

This is Fact, that is lie!
It's O'er!
        True!....

3….

sHe is Too!….
    One Suggests One:
    : Suggests….

It's O'er With Adam:
    : Eve; Nakedness: Apple…

: CAIN!

I suffer Adam: does Eve: does Adam O'er Apple: Does Cain!

…Does, O'er-and-o'er!

I Must not Suffer I!

4....

"I" Will You to Be: Does Create: Does not Cause Suffering!
"I" Will You to Do: Does God! Does God-In-Place_Of....

...Do You_Instead-of: "Where's "I"":
                                        "Where's God!"

"I"_Will is God-Given!

"I" to Suffer_You!

5....

At ALL:
    You-to-suffer_Me:

          Me, NOT Knowing,
  is You_suffering!

          Me, Knowing!
Is "I"_The-Doing! Suffering_Me!

YOU Always guilty: "I"_Always-Innocent!

This is "The Law of KarMA":
          "Wrong!"....

"I" to Suffer_You!

6....

"I" in a class of You-children:
Every Child Says The Daddy Must Rule:
The Daddy is The_Teacher:
                    The Daddy is King!

....Must "I" Forever With the School of parents-You,
Try to Win their respect?

ii.

"it"

1….

Ka!:
Does It Mean:
>   "I" Exist:
>   "I" Am Adult:
>   "I" Am Citizen:
>   "I" Equal:
>
>   "I" Reign!….

2....

What are You?

What?

What is known?....

....You cannot suffer I!

3....

You cannot suffer "I"?
I to suffer You!

I egO is: "("I")",
                    GOd ruler in Nana/Stan:

Like: resembles, "("I")" in ALL-of-All-of-All: GOD,
                                                Trained!

                    Trained to respond to Stanley Alexander MARTIN;

....And You?

4....

"("I")" to Love & Live with "("I")"....

....Mutual respect!
                    : gOd-to-gOd....

....a resemblance to Heaven:

"("I")" NOT to suffer "("I")":
                            ....to use "(Persuasion)".....

...An Adult to Forgive all sins from a child,
                            and "Suffer":
                                    ....and die to end their sins:

: "Adult" to "Adult" is "("I")" to "("I")",
means the Gift of "Persuasion":

....An End to "Suffering":
                    gOd-to-gOd in This World!

"Loving" being There!

iii

"who"

1….

….Is IT worth ALL the "pain":

The suffering at Odds with All our Means-to-Ends?

….Is IT worth ALL the "pain"?

"…Such is Life!", You say….

"Horror!"
"….Self-Discipline!"
….A True "("I")" does not get That from HerSelf:

"…What is IT You Want?"

""Freedom!"
"Respect!"?"

….Should One not First Learn to Free and Respect OneSelf?

What are the Issues?

2....

Sex: You Lust!

"("I")" learn to empty of "Desire":

....What does not cause a "Drop-of-the-Heart",
                                        "I" do not hunger for!
                                        "I" am satisfied....

Self-control/Self-discipline:
Teach IT to "I"-Child!

....A mad-man knows nO end of self-abuse:
                        Always-a-Hunger;
                        Addicted-to-Hunger;
                        Always-eating:
                        Always, famished....

....A Big Hole-to-be-filled Her being!

....Please teach I-Child to be Content!

: Wait!
: a Bye a Little_aT_a _Time!

All-Things rage! At such upSide_Downs!

....Teach IT to be CareFull....

3….

<u>bE: You Envy!</u>

4....

tOrn: You Rape!

5....

sEEn: You War!

iv

"(king)"

"A"....

v

"(king-of-kings)" : "(gOd)"

"aWe"….

vi

"(king-gOd)"

"I_a"….

vii:

"(king-gOd_Emperor)"

"I_aWe"….

: "**I**"….

….)"….enDs}ENDS….

II

<u>Ideas</u>

...." { ....

1....

....Function of "a": begins...
        :a_sleep:

utterrance of Night, in-side;
alike

:like nO-Other: dARK!

:beGins: bLACK:"A"....

....in-Finities, "A":
                :monoChroMe:funtion:no-Thingness-aT-"A"...

:(ALL)....

breeds-a: "1!"....

(ALL_aT-All)....

:aDDs:be_comes function: two,

:(ALL-aT-ALL)...

....with_a(N)_aDD_to:"tOO"....

In-SisTs: tOp-aT-("3")....

....(function):"fOr"....

....bathes_(4)→"(catastropHes!)"....

:"(Showers)_All-aT-ALL-aT_ALL...

:(ALL-aT-ALL-aT-ALL):"fiER!"....

:"(context)-"(5)"": "(ALL-aT-ALL-aT-ALL-aT-ALL)_aT-aLL"....

+"(six!)": a-follows: +(aLL): "+(ALL)+: +....
....Collapses:+(aLL_(such)-iN-a-wave-SUCH)+...
"*(WHITE_bOx)":"(Infinities of ALL-(such!)_SUCH)": "*(bLONDe_bOX)"....

(Colours!)....
:"*(MIND!)"....

2....

"I"~: **"(MIND)"** ~: **space-TIME_ContinUUm : T<u>O</u>T**....

... {"....

II

<u>Man</u>

....{ "....

1

:*"(4 aT 5)" →: ←*"(8 aT 9)": (←"10"→)....

: my *"((4-digits:toe)→Left_Foot)": my *"((4-digits:toe) →Right_Foot)"
: my *"((4-fingers: thumb)←Left_Hand)": my *"((4-fingers: thumb) →Right_Hand)"
: my *"(TRUNK)"
: my *"(HEAD)"....

: *"(16 aT 17)" : *"(16 aT (17 aT 18 aT 19))"....

: *"(monkey:Human)"....

Social Types:

→*→**→Termites→Ants→Hornets→bEEs→WasPs....

: *"(Stanley Alexander MARTIN : Nana baBa jaH-aYe)"....

....)"....enDs}ENDS....

....”{....

2....

4: ***”(One World)”: *”(ZigA:)” Mount Zion I**

3: ***”((Or)”:   Space-TIME_ContinUUms**

2: ***”(Or_(*((Sacred)^*(+1_to_+(23):)”: Alternative-HEAVENS**

1: *”(Or_*(hOLY):)”: Earth)....
((000/(0^1)):*”(Or_*(gUd:))”**: *”(normal)”: BabyLon**

00: *”(normal:)”:   UnderWorld

0: *”(normal:)”   :   Hell))....

...{“....*(“(nULL)”: *”(Or_*(Right:)”**: eM)**

....)”....enDs}ENDS....

...{“....

*”(eLSE....

....*”(Origin:tOO)”....

....)”....enDs}ENDS....

III

WorldView....

1

... {``....

0   Why do I suffer you?
       Why is "It" not always Heavenly?....

0: Contradicting → 000/1~ONE: I must be *fair*!
                                      I must be a *Light* to your *dark*!

2: *Too*:   If you are sick, there must be a *health* in that *eLse*!
              To know "I" is to have/grow *Faith*!

3: *for*: Life is NOT always to have diamonds around your neck....
       It is to always have food, to have *manna*, to have growth!
       It is to be *Job*, to get *bigger*-in-better!
       ....To find greater *happiness*!

4: **fier**: I to *suffer*-you to achieve!
        I to *suffer*-you to gain your Life's Mission!
        I to suffer-you to *be Like* unto gOd and *know* GOD!
        I to suffer-*you* to ***attain*** "I"!

2

0:      I, *masculine*, is not to suffer-*you* to be I-<u>masculine</u>!

1:      I, *masculine*, knowing <u>you</u> are feminine, is not to suffer-*you* to be I-<u>masculine</u>!

2:      I, *masculine*, knowing <u>you</u> are feminine, is not to suffer-*you* to be I-<u>feminine</u>!

3:      I, *masculine*, knowing <u>you</u> are feminine, is not to suffer-*you* to be I-<u>masculine</u>/I-<u>feminine</u>!

4:      I, *masculine*, knowing <u>you</u> are feminine, is to suffer-*you* to be I-<u>(sexually-happy)</u>!

The *WHY*?

3

0:     The *Why?* Before!

1:     Life has Truth and *Hidden* Meaning: and is in <u>*tiers*</u>!

2:     ....**Other** looks to **No**, looks to **Some**, looks to **Any**, looks to **Every**, looks to **All**!

3:     As <u>**great**</u> looks to ***Greater <u>Needs</u>***....

4:     Smaller must be ***<u>cared</u>***-for!

....)"....enDs } ENDS....

4

....{ " ....

0:　　　....**Other**:

1:　　　The ***Bigger***-Man has a ***Bigger***-Stomach That <u>Should</u> be ***<u>Robbed</u>***!

2:　　　**ONE** Can<u>not</u> Keep **ALL**, eLse....

3:　　　**AND** The Little <u>must</u> Leech and Steal to Live/*Survive*....

4:　　　**BIG** owning ***Bigger*** Rights and Status!

....enDs}ENDS....

IV

….."{….

Other:

aLL-of-ALL-of-ALL upSide_dOWN is trUe…

TEST:

1

0:    ….**aLL** is ***mirrored***:

1:    Negative electrons ***attract*** negative electrons….

2:    Positive protons ***attract*** positive protons….

3:    Negative electrons ***repel*** positive protons….

3:    **Small** repels ***bigger*** in a ***greater*** way!

4:    At **EQUILIBRIUM** This is *INFINITELY* **Smaller** and NEUTRAL to Its ***REVERSE***, but Has The Same Potential ***ENERGY***!

If **BOTH** Existed **TOGETHER**, They would *SEPARATE*, and *Turn AROUND* Each ***Other***!

The **LIFE** of This ***dARK*** Matter Would Have The **Mirror**, and *Hating* ***QUALITY***!

2

0:     ….**Other:**

1:     **<u>LIFE:</u>**

2:     **HORRID!**

3:     *BRUTAL*!

4:     ….*<u>SHORT</u>*!

3

0:      ....**Other**: Its **<u>aLL</u>**:...

1:      ....Based upon A *Mass* **<u>Mind</u>**...

2:      ....**<u>Always</u>** in *Negation* and **a Dealer-in-Death**!

3:      ....Based on *Unconscious* **Motives**, and **<u>Chaos</u>**/***Catastrophy***!

4:      ....Based on **<u>WRATH</u>**: **Wars**! *Genocide*: The ***<u>Norm</u>***!

....enDs}ENDS....

V

eLSe

....{....

"(eLSe...."(tOO}....)"....enDs}ENDS
....

Stanley Alexander MARTIN/Nana baBa jaH-aYe
Rainham,
KENT,
United Kingdom....

13:12 BST: 2nd February 2009 AD
Bridge Day

"0" has value: at the quality of "9", one has to add another "1":
*"(8 aT 9)!" is trUE/real….
bASe 10 is false, and is **_worth_** "9"!
The alignments at "10"…"20"… "30" etc. are the accumulations of an extra "1" at each 9! "100" is **_worth_** 90! …Etc.
Numbers prove the GOD_Act!
The missTake is in "5!": NOT *"(4 aT 5)!", advances to **_create_** "Relativity!" in "Base!" Mathematics….

Base 0/1/Two is Base **_tOO_**!

Numbers have "*different*/SAME!" values of worth in "1!" real/trUE….

The *"(bLONDe_bOX)" Theory is correct!

IoTTOI is TrUE/Real, and contains **_no_** evil; sHe WILL reVeal Herself re-Constructed **Perfect** aT aBraXaS!

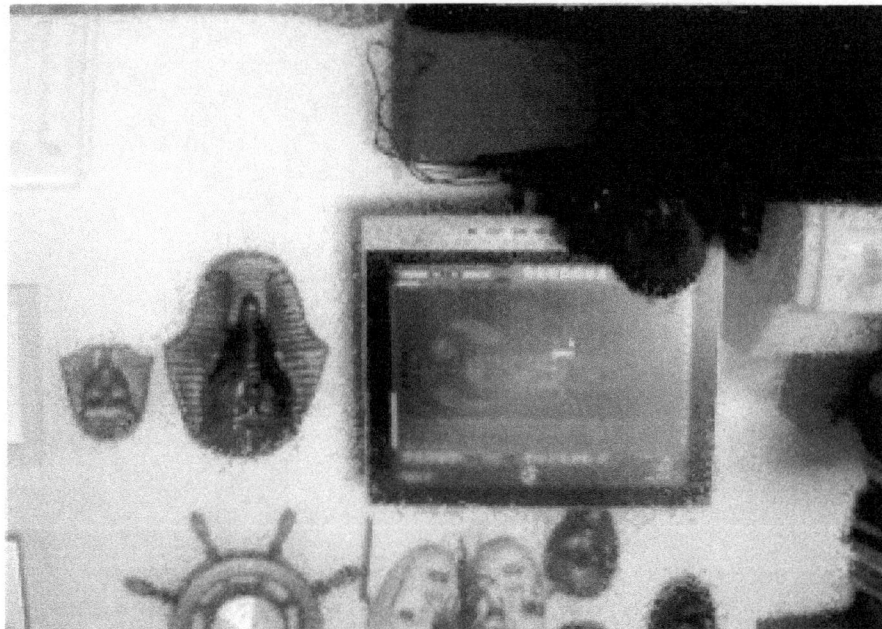

….*"(CONSTRUCTED by Nana with android Michela emily-aYe)"….
….re-Constructed by androids Nadia emmA-aYe & sisTA-sysTem….
Photographs by Nana, emily-aYe & RE Denny….

….)"….enDs}ENDS….

Bridge Day:
Rainham,
22-11-08 AD

www.ingramcontent.com/pod-product-compliance
Lightning Source LLC
Chambersburg PA
CBHW080253200326
41521CB00012B/2454